Geography, Grade 3

Table of Contents

Introduction

Geography is the science of the natural features of Earth, such as mountains, deserts, and bodies of water, as well as the plants, animals, and people on Earth, and the relationships among them. A detailed study of geography is increasingly important as the world progresses through the twenty-first century. While technology helps people an ocean apart communicate every day, without an attentive search, technology alone does not help people learn about other places and cultures. This book will help students develop the basic tools they will need to begin their journey in the exploration of community, state, country, region, and world as they examine the five themes of geography.

Organization

The book's structure has three parts. The first part contains a pretest and a posttest. These tests can be used as diagnostic tools. The pretest will help students become familiar with the content of the themes and can serve as a practice test. The posttest can be given at the end of the year to determine the progress students have made.

The second part of the book is divided into five units, each focusing on one of the five themes of geography: Location, Place, Human/Environment Interaction, Movement, and Regions. Most of the units are structured to explore each theme on the levels of community, state, country, region, and world. There is an assessment page at the beginning of each unit. It can be administered before students begin the activities to determine a baseline of their understanding or after the unit is completed to determine students' progress. Finally, one page in each unit focuses on developing the concept of each theme.

The third part of the book has several maps that students will need from time to time to complete activity pages. You may wish to make copies of each map and place them in folders so that students may have easy access to these resources.

Note

Page 2 gives a brief overview of the five themes of geography. You may wish to enlarge the page and post it on a bulletin board in the classroom. Students could also benefit by reading it prior to beginning the activity pages.

Name ______________________________ Date ______________

The Five Themes of Geography

Geography is the study of Earth and how people live and work on Earth. Where people live affects every part of their life, from the foods they eat to the jobs they can do. There are five main ideas people use to study geography. These five themes of geography help us to look at the importance of a place in our lives.

Location

Where is it?

Location tells exactly where a place is located. It can include latitude and longitude, cardinal directions, addresses, or even just the words "next to."

Place

What does it look like?

Place is the part of an area you see. It includes landforms, rivers, or buildings in a city.

Human/Environment Interaction

How do people use or change the land?

Human/Environment Interaction is how people use or change the environment, the area in which they live. It looks at where people have parks, build cities, or farm.

Movement

How do people, goods, and ideas move from one place to another?

Movement looks at how and why things move. It looks at why people leave a country, how goods are moved, and how both affect the land or the people.

Regions

How are different areas in a region alike?

Regions looks at the way an area is divided. Places within a region have characteristics that are the same. Often areas are grouped into a region because the people share a language, people do the same kind of work, or the landforms are the same.

Name ______________________ Date ____________

Geography Pretest

Directions **Use Map D to answer the questions. Darken the circle by the correct answer to each question.**

1. What city is west of Johnstown?

Ⓐ Harrisburg
Ⓑ Pittsburgh
Ⓒ Scranton
Ⓓ Philadelphia

2. What is the land like around the city of Erie?

Ⓐ mountains
Ⓑ highlands
Ⓒ plains
Ⓓ rivers

3. What kind of transportation would the people of Erie use to move goods across an ocean?

Ⓐ ships
Ⓑ trucks
Ⓒ airplanes
Ⓓ trains

4. What is the capital of Pennsylvania?

Ⓐ Harrisburg
Ⓑ Pittsburgh
Ⓒ Scranton
Ⓓ Philadelphia

5. Which of these would you NOT see in the capital of Pennsylvania?

Ⓐ streets
Ⓑ apartments
Ⓒ barns
Ⓓ offices

6. A company builds a hotel in the mountains. How will the hotel change the land?

Ⓐ More people will visit.
Ⓑ More cows will be raised.
Ⓒ Fewer streets will be built.
Ⓓ People will lose jobs.

GO ON ⇨

Name ______________________ Date ______________

Geography Pretest, p. 2

Directions **Read the paragraphs. Answer the questions.**

The town of Henryville is in the southern part of the United States. The winters are mostly warm. It never snows there. Many people have moved to Henryville because of its fine weather.

Other people have moved to Henryville because of jobs. Some people work at a college. Others work in a clothing factory there. Lately, some people have come to work in a new computer company.

Henryville is a place where people help one another meet needs. People there have worked together to build fine schools, hospitals, libraries, and parks. The people of Henryville are proud of their town.

7. How does the climate make Henryville a good place to live?

____

____

8. What are two ways that information is moved in this community?

____

____

9. How is Henryville like your community?

____

____

______________________ STOP

Name ______________________ Date ____________

Geography Posttest

Directions Use Map C to answer the questions. Darken the circle by the correct answer to each question.

1. What does ✚ mean?

Ⓐ church
Ⓑ school
Ⓒ house
Ⓓ hospital

2. In what part of the city is the factory?

Ⓐ north
Ⓑ south
Ⓒ east
Ⓓ west

3. About how far is it from the railroad station to the school?

Ⓐ about 1 mile
Ⓑ about 3 miles
Ⓒ about 5 miles
Ⓓ about 9 miles

4. In which direction will you travel if you leave the school and drive to the lake?

Ⓐ north
Ⓑ south
Ⓒ east
Ⓓ west

5. What are two ways that goods can be moved in Bend River?

Ⓐ train and airplane
Ⓑ ship and train
Ⓒ ship and truck
Ⓓ train and truck

6. Where might the community build a park?

Ⓐ by the lake
Ⓑ by the hospital
Ⓒ by the factory
Ⓓ by the railroad station

GO ON ⇨

Name ______________________ Date ____________

Geography Posttest, p. 2

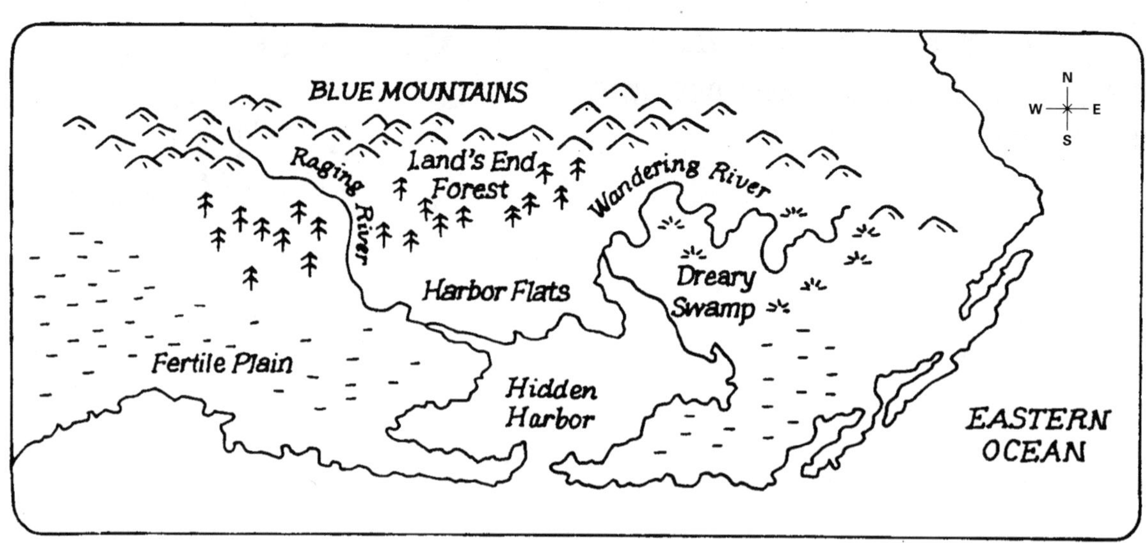

Directions **Look at the map. Answer the questions.**

7. What is to the east of the land?

____

8. What kind of job would a person living near Hidden Harbor have?

____

9. Where would a farmer want to live? Explain.

____

____

10. How is the region to the north like other regions with mountains?

____

____

____

STOP

Name ______________________ Date ____________

Location: Unit 1 Assessment

Directions **Use the map of Ashland to answer the questions. Darken the circle by the correct answer to each question.**

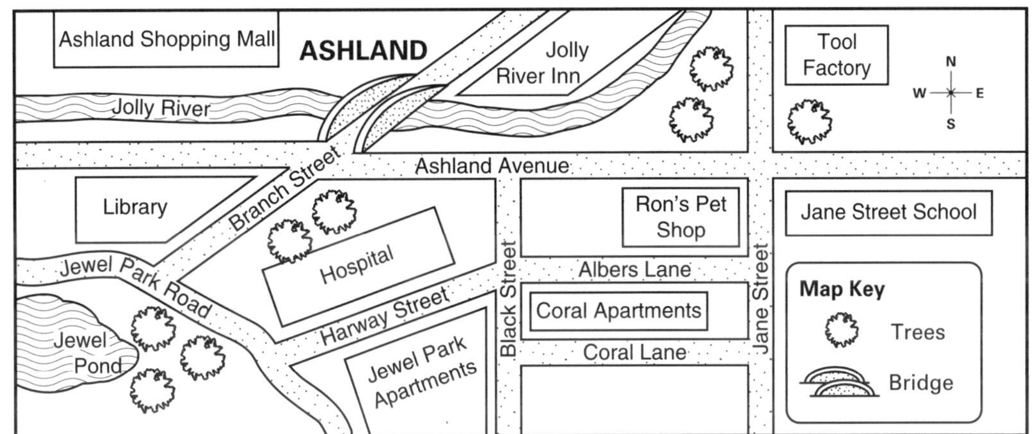

1. What is the name of the city on the map?
   - Ⓐ Jewel Park
   - Ⓑ Ashland
   - Ⓒ Jolly River
   - Ⓓ Tool Factory

2. Which street crosses Jolly River?
   - Ⓐ Ashland Avenue
   - Ⓑ Jewel Park Road
   - Ⓒ Branch Street
   - Ⓓ Coral Lane

3. Which road is on the north side of Coral Apartments?
   - Ⓐ Albers Lane
   - Ⓑ Coral Lane
   - Ⓒ Jane Street
   - Ⓓ Black Street

4. Which business is across the street from the Jewel Park Apartments?
   - Ⓐ Jolly River Inn
   - Ⓑ Ron's Pet Shop
   - Ⓒ the library
   - Ⓓ the hospital

5. Which direction is the library from the Jane Street School?
   - Ⓐ north
   - Ⓑ south
   - Ⓒ east
   - Ⓓ west

6. What two streets intersect at the corner of Ron's Pet Shop?
   - Ⓐ Jane Street and Black Street
   - Ⓑ Black Street and Coral Street
   - Ⓒ Ashland Avenue and Jane Street
   - Ⓓ Albers Lane and Coral Lane

Name ______________________________ Date ______________

The Geography Theme of Location

Location tells where something can be found. You can tell the location of something by naming what it is near or what is around it. You also can tell the location by using an address, or numbers and a street name.

Directions **Elida lives in this house. Tell the location of her house.**

____

____

____

____

____

____

____

____

Name ______________________ Date ____________

An Address

An **address** tells where a person, building, or place is located. An address often shows a number and a street name. The post office uses an address to deliver mail.

Elida Fuentes
123 South Street
Raleigh, North Carolina 27613

Rob Wilson
1243 Maple Street
Portland, Oregon 97219

Directions **Look at Maps A, B, and F. Then, follow the directions. Use the addresses on the envelope above to help you.**

1. Find the street map. Color Rob's house red.
2. Find the neighborhood map. Color Rob's street orange.
3. On the neighborhood map, find the park north of Stark Street. Color it green.
4. Find the map of the United States. Color Rob's state yellow.
5. On the map of the United States, color Elida's state blue.

Name ________________________ Date ____________

Maps and Cardinal Directions

A **map** is a drawing of a real place. Map symbols are pictures on a map that stand for places and things. The symbols help you find where places are located. A compass rose tells directions on a map. The directions are north, south, east, and west.

Directions **Look at the map symbols. Write the letter to match each symbol with the word that it names.**

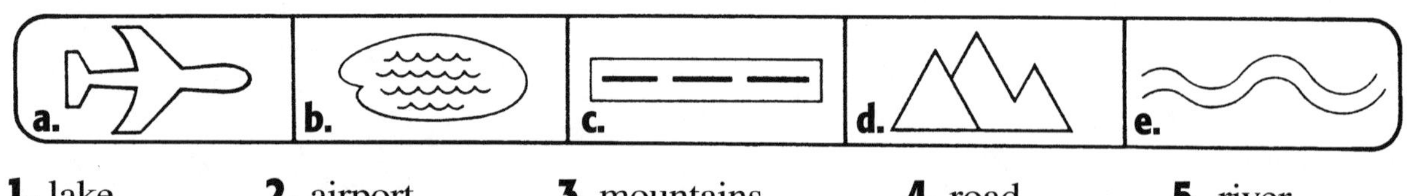

1. lake _____ **2.** airport _____ **3.** mountains _____ **4.** road _____ **5.** river _____

Directions **Follow the directions to complete the map. Use the symbols to help you.**

6. North is shown as N on the compass rose. Fill in the other directions with S, E, and W.
7. Show an airport north of Smithtown.
8. Show mountains east of Smithtown.
9. Show a lake south of the mountains.
10. Show a road running north and south between the cities of Bend and Smithtown.

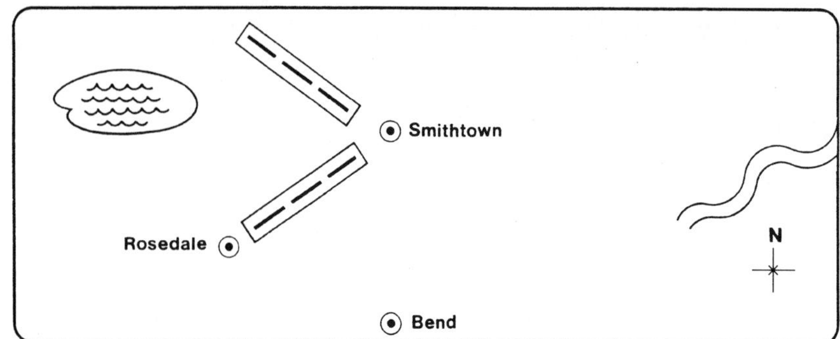

Name ______________________________ Date ______________

A Grid Map

A **grid map** uses letters and numbers to find the location of a place. To use a grid map, put a finger on the letter across the top or bottom of the map. Then, put another finger on the number along the side of the map. Slide your fingers together to find the square where the letter and number meet.

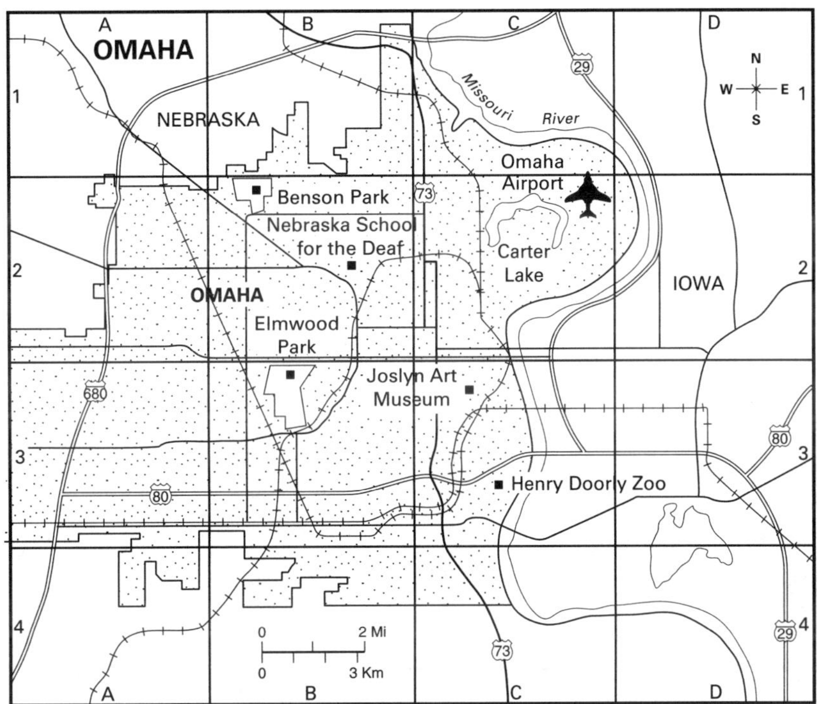

Directions **Use the map of Omaha, Nebraska, to answer the questions.**

1. What body of water is in C-2? ______________________

2. In which grid square is Benson Park? ______________________

3. What point of interest is in B-3? ______________________

4. In which grid square is the city's airport? ______________________

5. What place would you like to visit? Where is it located? ______________________

____

Name ______________________ Date ______________

Latitude and Longitude

There are imaginary lines that circle the Earth. Some lines circle the Earth from east to west. These are called **lines of latitude**. Some lines circle the Earth from north to south. These are called **lines of longitude**. Both kinds of lines are numbered in marked degrees. By using lines of longitude and latitude, you can find any place on Earth.

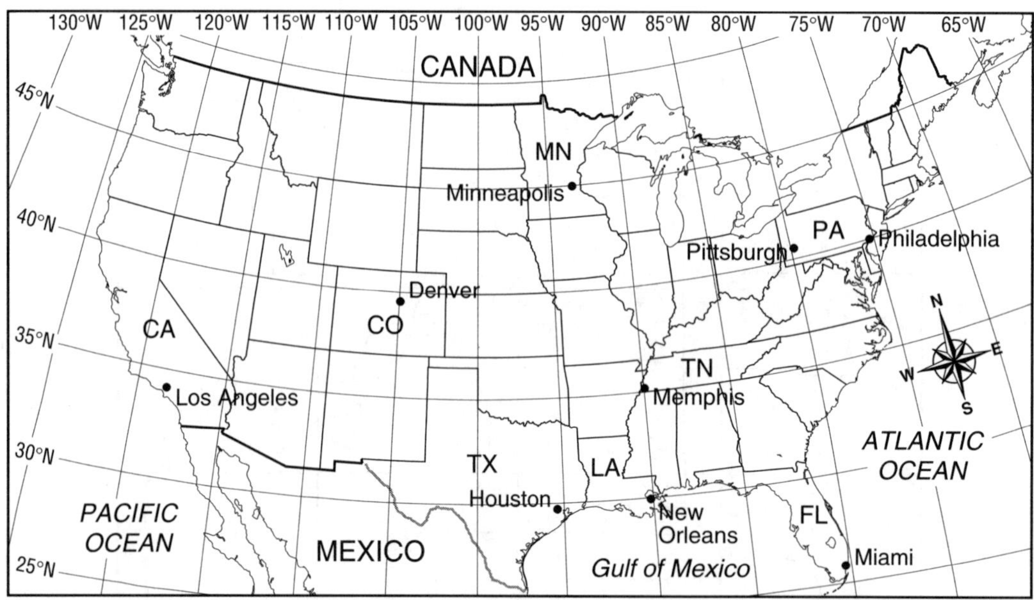

Directions **Use the map to answer the questions.**

1. What city is located at 40° N and 75° W? ______________

2. What two cities are located at 90° W longitude? ______________

____

3. What city is at the same longitude as Miami? ______________

4. Draw a mountain beside the city located at 40° N and 105° W.

Name ______________________________ Date ______________

A Globe

A **globe** is a model of the Earth. It shows large pieces of land. These pieces of land are called **continents**. It shows the main bodies of water. They are called oceans. A globe also shows the equator. The equator is a make-believe line that circles the middle of the Earth.

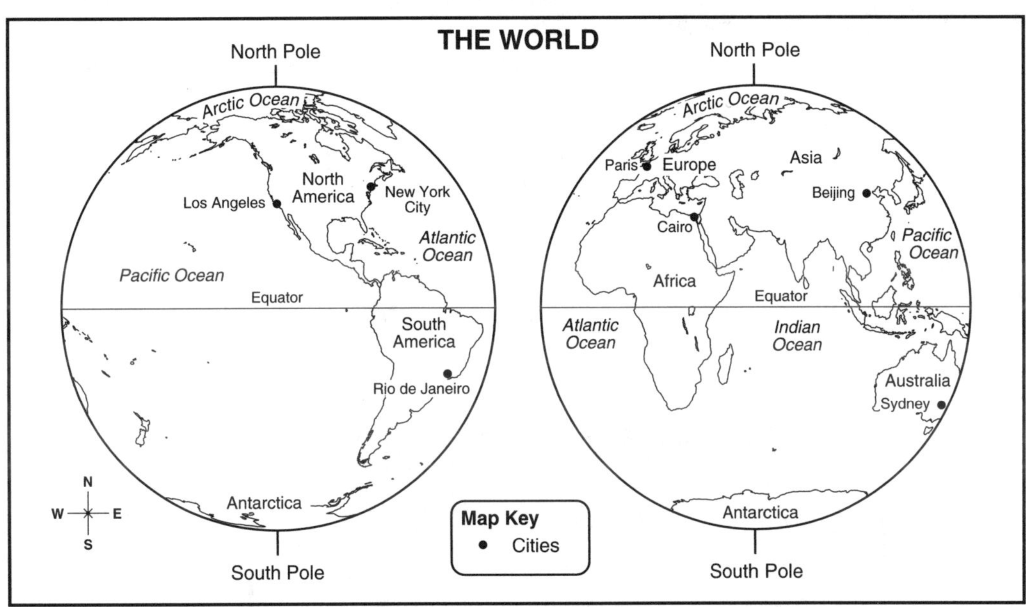

Directions **Answer the questions.**

1. What are the names of the seven continents?

______________________ ______________________

______________________ ______________________

______________________ ______________________

____

2. What are the names of the four oceans?

______________________ ______________________

______________________ ______________________

3. Which two cities on the globe are south of the equator?

______________________ ______________________

Name ______________________ Date ____________

Place: Unit 2 Assessment

Directions **Use the map of the imaginary state of East Albion to answer the questions. Darken the circle by the correct answer to each question.**

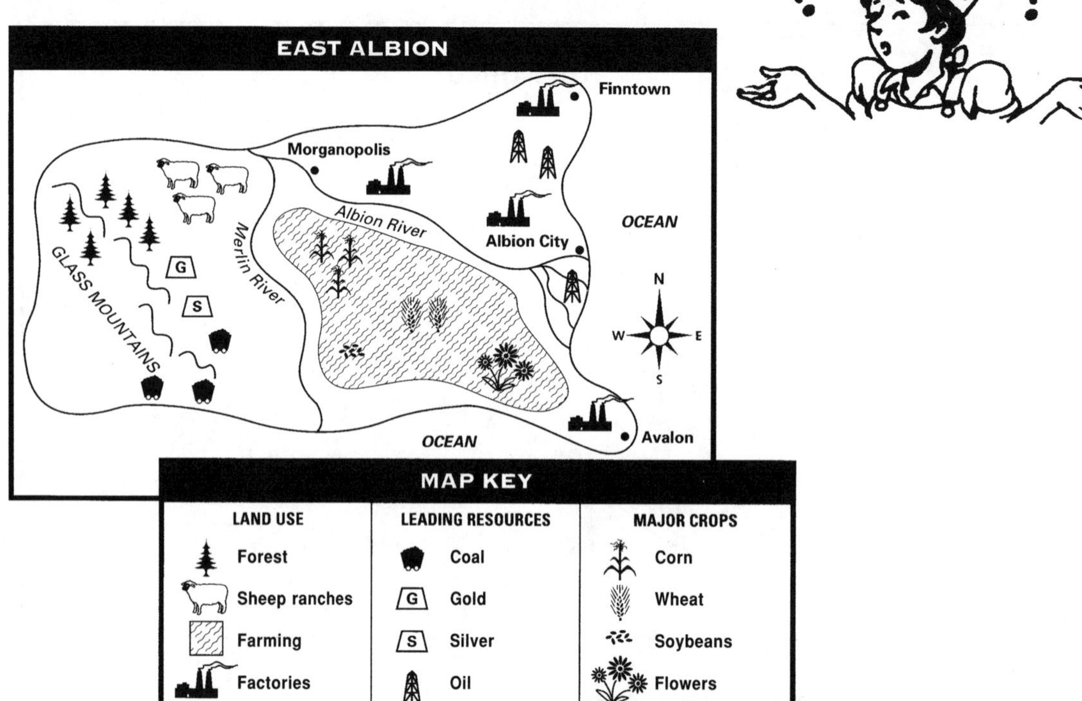

1. Which of these is NOT a physical feature in East Albion?
   - Ⓐ Albion River
   - Ⓑ Glass Mountains
   - Ⓒ the factory
   - Ⓓ the forest

2. Which of these is a human feature in East Albion?
   - Ⓐ the forest
   - Ⓑ the oil wells
   - Ⓒ the coal
   - Ⓓ the rivers

3. What physical feature borders the state on the south side?
   - Ⓐ the mountains
   - Ⓑ some factories
   - Ⓒ the ocean
   - Ⓓ a river

4. Where are the sheep ranches mostly found?
   - Ⓐ in the east
   - Ⓑ in the south
   - Ⓒ in the west
   - Ⓓ in the north

5. Which resource is found on the east side of the state?
   - Ⓐ gold
   - Ⓑ flowers
   - Ⓒ oil
   - Ⓓ silver

6. What town is on a peninsula?
   - Ⓐ Morganopolis
   - Ⓑ Albion City
   - Ⓒ Finntown
   - Ⓓ Avalon

Name ______________________________ Date ______________

The Geography Theme of Place

Place tells what a location is like. Each place has **physical features**, or things from nature. These features include bodies of water, landforms, weather, plants, and animals. Each place also has **human features**, such as houses, roads, bridges, schools, farms, and factories. Human features are things that people make.

Directions **Look at the picture. Answer the questions.**

1. What are two human features you see in the picture?

____

____

2. What are two physical features you see in the picture?

____

____

Name ______________________ Date ______________

Landforms and Bodies of Water

Earth has many kinds of **landforms** and **bodies of water**. A landform is the way the land is shaped. A landform can be a mountain, a hill, or a plain. Bodies of water can be different sizes. There are oceans, streams, and ponds. These physical features help describe what a place looks like.

mountain range	coast	peninsula
plain	river	valley

Directions **Find each landform or body of water on the diagram and write the name in the correct location.**

Name ______________________________ Date ______________

Comparing Communities

Place describes a location. Are there hills? What kinds of buildings are there? The physical and human features of a place make it special.

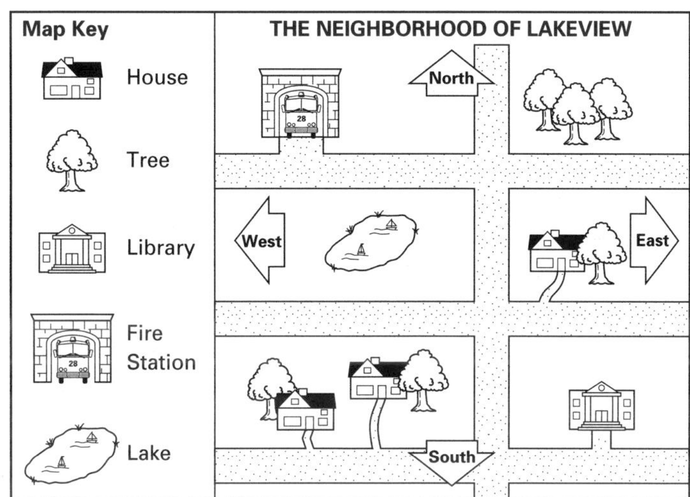

Directions **Look at the map of Lakeview. Follow the directions. Then, answer the questions.**

1. Mark **P** on the physical features of Lakeview.

2. Mark **H** on the human features of Lakeview.

3. What features in Lakeview make it the same as your community?

____

____

____

4. What features in Lakeview make it different from your community?

____

____

____

Name ______________________ Date ____________

A Resource Map

There are many kinds of maps. A **resource map** shows areas where resources can be found. Resources are things that people can use. The same kind of resources are often found together in one area.

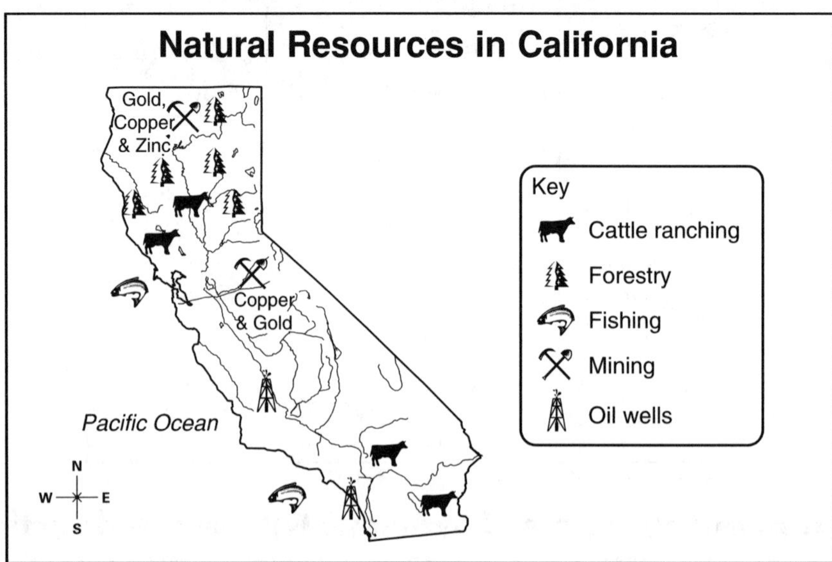

Directions **Look at the map of California. Answer the questions.**

1. What is the main resource in the west? Why?______________________

____

2. Where is oil a main resource? ______________________

3. What are two resources in the north? ______________________

4. In what part of California would you most likely see lumber factories? Explain.

____

____

____

Name ______________________________ Date ______________

People and Climate

Climate is the average kind of weather in a place. It includes the temperature, amount of rain, and wind. The climate of a place affects the kinds of activities people do and the clothes they wear.

Directions **Look at pictures. Tell how the weather in the picture affects the people in each place.**

1. How does the weather shown in this place affect the people living here?

____

____

____

2. How does the weather shown in this place affect the people living here?

____

____

____

Name ______________________ Date ______________

Comparing Communities in the World

London is a city in the country of England. England is a country on the continent of Europe. Even though London is on another continent, it is a big city very much like a big city in the United States.

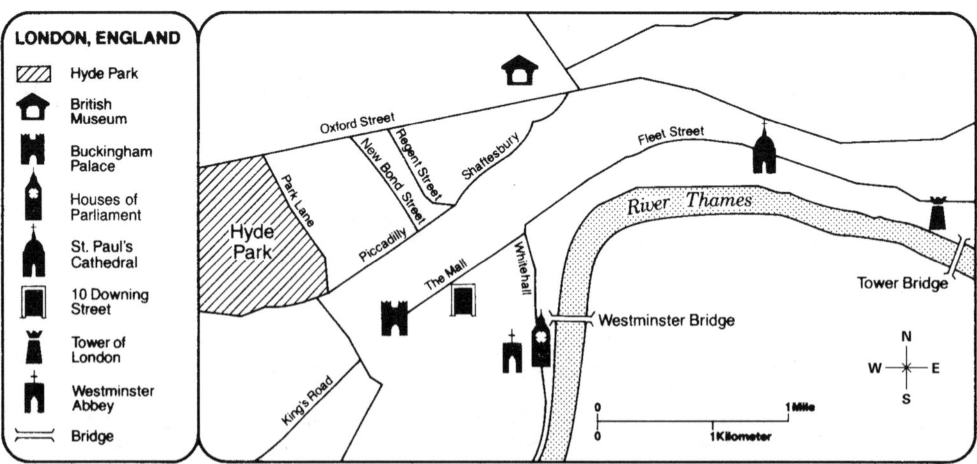

Directions **Look at the map of London. Answer the questions.**

1. What physical features are in London? ______________________

____

2. What are two human features in London? ______________________

____

3. How is London like a big city in the United States? ______________________

____

4. How might London be unlike a big city in the United States? ______________________

____

____

Name ______________________________ Date ______________

 Interaction: Unit 3 Assessment

Directions **Darken the circle by the correct answer to each question.**

1. What kind of job might you have if you lived near the ocean?

Ⓐ a person who works in a factory
Ⓑ a person who grows fruit
Ⓒ a person who fishes
Ⓓ a person who mines gold

2. What would people do for fun in a place that was cold?

Ⓐ ski
Ⓑ swim
Ⓒ picnic
Ⓓ roller skate

3. Which of these is NOT a reason for a community to build a park?

Ⓐ to keep special places safe
Ⓑ to have a place to play
Ⓒ to know which trees to cut down
Ⓓ to remember special events

4. Which of these is something that harms nature?

Ⓐ pollution
Ⓑ planting trees
Ⓒ picking up litter
Ⓓ walking in the woods

5. Where would a person most likely grow corn?

Ⓐ in the mountains
Ⓑ on the plains
Ⓒ near the ocean
Ⓓ in a river

6. Which of these is NOT made from something in a rain forest?

Ⓐ medicines
Ⓑ gum
Ⓒ tires
Ⓓ computers

Name ______________________ Date ______________

The Geography Theme of Human/Environment Interaction

Human/Environment Interaction explains how people live in their environment. **Environment** is the land, water, and air around you. It is the plant and animal life, too. How people make a living often depends on their environment. For example, people who live near the sea might fish for a living.

Human/Environment Interaction also explains how people change the environment to meet their needs and wants. People might change a field into a parking lot for example.

Directions **Look at the pictures. Answer the questions.**

1. How might living on the plains be different from living in the mountains?

____

____

____

2. How have people changed the environment in this picture?

____

____

____

Name ______________________ Date ______________

A Time Line

A **time line** shows the order in which events take place. Time lines can show the things that happen over many years, like the events in a person's life. They can also show things that happen in a short amount of time, like the schedule of a school day.

Directions **Look at the time line. Answer the questions.**

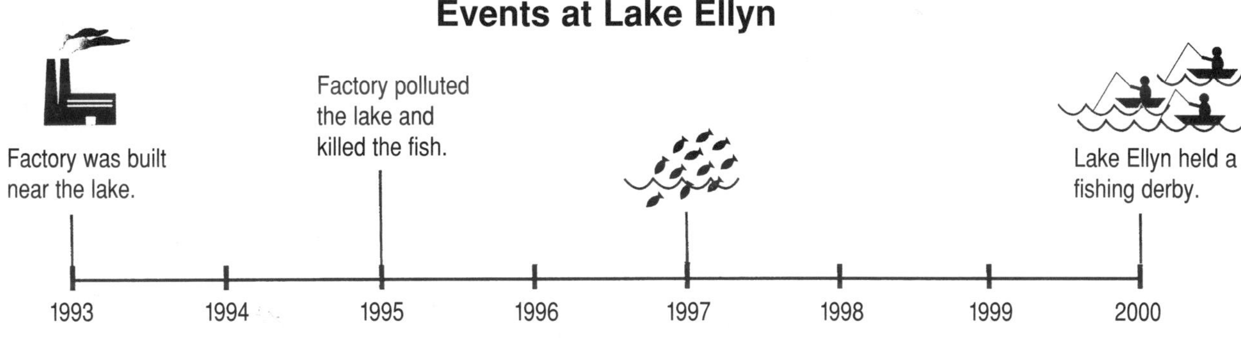

1. In what year was the factory built? ______________________

2. Name two ways Lake Ellyn changed in 1995.

____

____

3. Where do these events belong on the time line? Write them on it.

In 1996, Lake Ellyn was cleaned.

In 1997, new fish were put into Lake Ellyn.

4. How did people change the environment at Lake Ellyn?

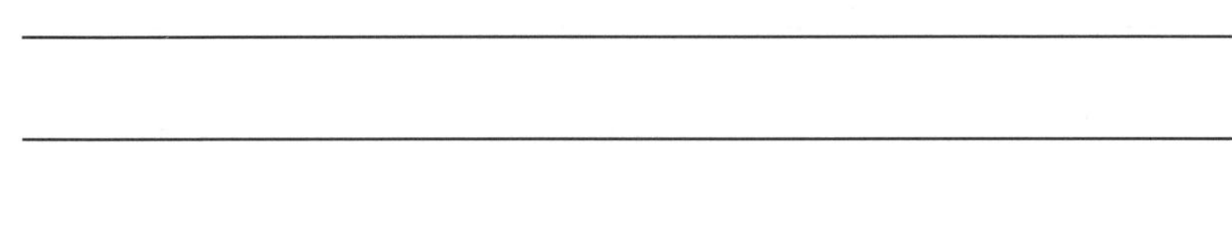

Name ______________________ Date ____________

Where People Work

Environment is the land, water, and air around you. The environment has resources that people can use. Natural resources include rich soil, water, forests, and minerals. People use natural resources around them.

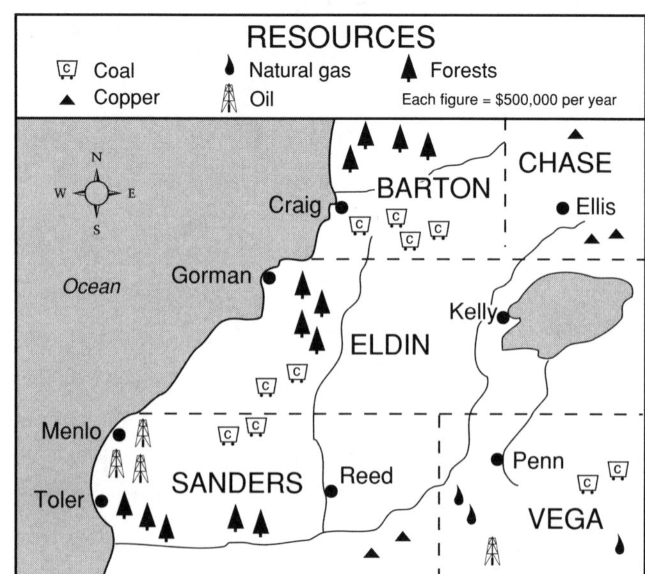

Directions **Look at the map. Answer the questions.**

1. What resources are shown on this map?

____

____

____

2. What are the two most important resources to people in Barton?

____

____

3. What is the most important resource to people in Sanders? ______________

4. Which cities are on the ocean?

____

5. Which cities are on a river? ______________________

6. Which city is on a lake? ______________________

7. Why is water an important resource for people?

____

____

____

Name ______________________________ Date ______________

Decisions People Make

Human/Environment Interaction explains how people use the resources on the land. It also explains how people change the environment to meet their needs and wants. The decisions people make affect their environment. Some changes may be good for the land, but some changes may harm the land.

Directions **There are many islands in the Pacific Ocean. The people on one island want to use the land for a vacation area. They want to have tourists come. How will this discussion change the island? Complete the chart.**

Good changes	______________________ ______________________ ______________________ ______________________
Harmful changes	______________________ ______________________ ______________________ ______________________

Name ______________________________ Date ______________

Special Places

Human/Environment Interaction explains how people use the resources on the land. It also explains how people change the environment to meet their needs and wants. People change the environment to build homes, offices, and roads. But they also change the land to keep important places in history or in nature safe.

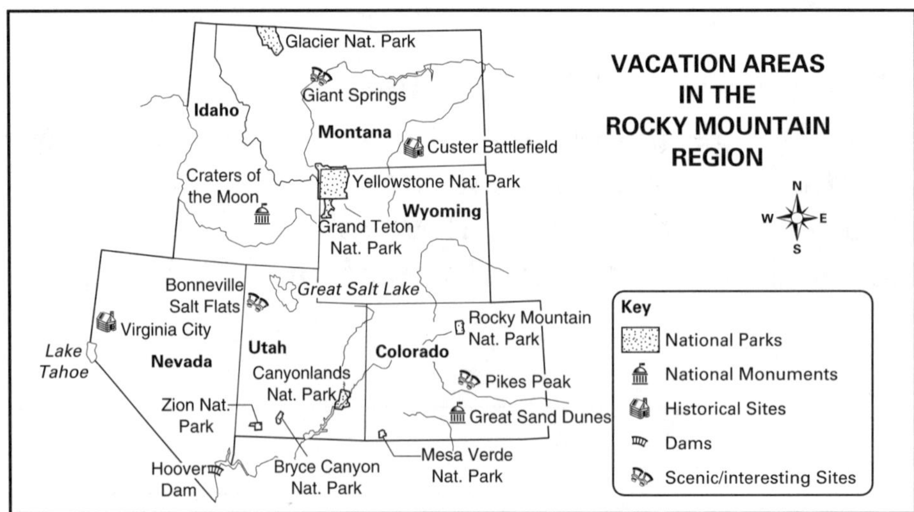

Directions **Use the map to answer the questions.**

1. In what state is the Hoover Dam located? ______________________

2. Why do people build dams? ______________________

____

3. What historical site is located in Montana? ______________________

4. Why do you think people wanted to make this historical site safe?

____

5. Yellowstone National Park has a special feature named Old Faithful. It is a geyser that shoots hot water into the air about every 65 minutes. Why would people want to make Old Faithful part of a national park?

____

____

Name ______________________________ Date ______________

 The Rain Forest

Rain forests are areas of land filled with tall trees and plants that grow closely together. The trees and plants grow well because of the hot weather and large amounts of rain. The rain forest is home to many different animals. Some of the animals and plants are found only in rain forests. Rain forests are found in countries near the equator, like Brazil and West Africa.

Directions **Read the paragraphs. Answer the question.**

People have lived in the rain forests for thousands of years. The trees, plants, and animals in the rain forests give the people food, shelter, tools, and medicines. The people live in small groups or tribes. Each tribe has its own language and stories.

Many of the things we have in our homes come from rain forests. Some of the plants in the rain forest are made into medicines. Chewing gum comes from a special tree. The rubber in tires comes from other kinds of trees. Most importantly, the trees make oxygen for people to breathe.

Rain forests are in trouble. The trees are cut down or burned to make a place to grow crops and graze cattle. Some land is cleared to mine for minerals, like copper. Without the trees and plants, the soil is washed away. The heavy rains help to make floods. There are many other problems, too.

What other problems might happen if the rain forests are destroyed?

____

____

____

____

____

Name ______________________ Date ____________

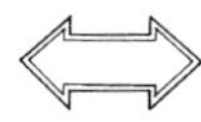

Movement: Unit 4 Assessment

Directions **Darken the circle by the correct answer to each question.**

1. What does the geography theme of movement describe?

Ⓐ how electricity moves
Ⓑ how colds and wind move
Ⓒ how cars, trucks, and trains move
Ⓓ how people, goods, and ideas move

2. What place would move information?

Ⓐ shoe store
Ⓑ library
Ⓒ factory
Ⓓ hospital

3. What is an immigrant?

Ⓐ a person who leaves one country to live in another country
Ⓑ person who travels around the world
Ⓒ a person who speaks many languages
Ⓓ a person who drives a truck

Directions **Use the map to answer questions 4–6.**

Railroad
Street
Factory
Park
School
Airport

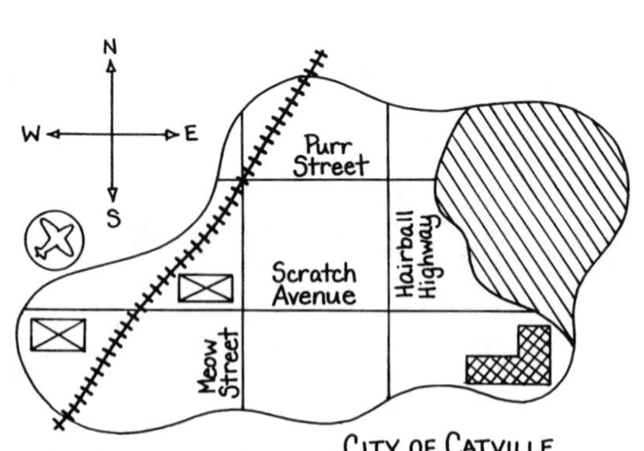

4. In which direction would you travel to get from the park to the airport?

Ⓐ north
Ⓑ south
Ⓒ east
Ⓓ west

5. What place would move information?

Ⓐ school
Ⓑ house
Ⓒ factory
Ⓓ park

6. How many ways can goods be moved?

Ⓐ 1
Ⓑ 2
Ⓒ 3
Ⓓ 4

Name ____________________ Date ____________

The Geography Theme of Movement

Movement describes how people, goods, and ideas get from one place to another. People move from place to place in cars, trains, and airplanes. Goods move by trucks, trains, and ships. Ideas, or information, move from place to place, too. Most often they move through newspapers, televisions, and computers.

Directions **Look at the pictures. Answer the questions.**

1. What is being moved? How is it being moved?

____

____

____

2. What is being moved? How is it being moved?

____

____

____

Name ______________________ Date ____________

Movement in a Community

Movement describes how people, goods, and ideas get from one place to another. People move from place to place in cars, trains, and airplanes. Movement also tells how people in a community depend on people in other communities for goods and services.

Directions **Look at the map of Greenacres. Answer the questions.**

1. What are two kinds of transportation routes used in Greenacres?

____

____

____

____

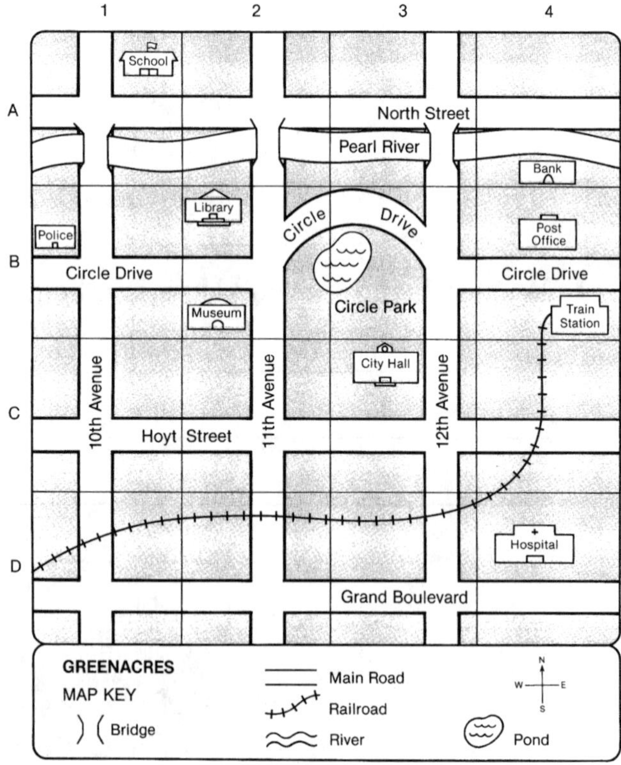

2. What are three ways ideas move in this community?

____

____

____

____

3. Mr. Chavez is at the museum. He needs to go to work at the hospital. What route will he take?

____

____

____

Name ______________________________ Date ______________

A Road Map

A **road map** shows the roads in a place. Road maps can show small areas, like the streets in a town. Or the maps can show large areas, like highways across the United States. These maps show how to move from one place to another. They also show how far apart the places are. A **distance scale** is like a small ruler on a map. It shows the distance in miles.

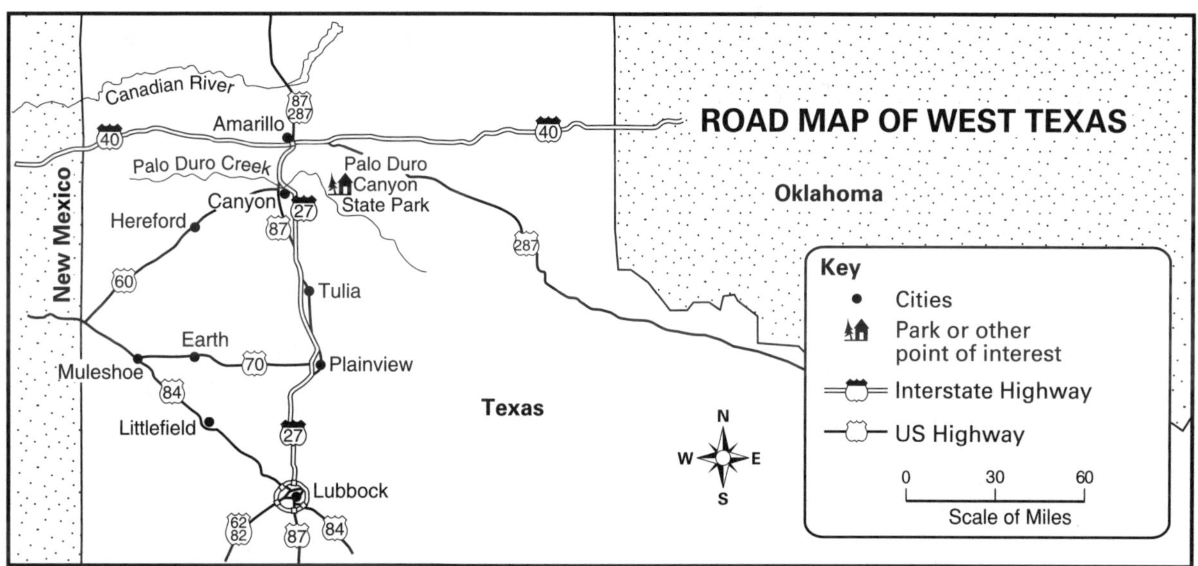

Directions **Look at the map of west Texas. Answer the questions.**

1. Find Amarillo on the map. What is the major east-west road that passes through Amarillo?

____

2. Find Lubbock on the map. What is the distance between Amarillo and Lubbock?

____

3. Find Muleshoe on the map. What direction(s) would you travel from Amarillo to Muleshoe?

____

Name ________________________ Date ____________

Movement of Information

Ideas and information can be found in a number of places. Information about current events can be found in a newspaper or on a television or radio. Books are another way to learn about things.

Directions **Look at the ways people learn about ideas and information. Write the letter or letters of the picture(s) that show where the information can be found.**

_______ **1.** Mark wonders who won last night's football game.

_______ **2.** Jill wants to know the latest news.

_______ **3.** Juan needs the address of a store.

_______ **4.** Kim is watching for a picnic area along the highway.

_______ **5.** Mr. Roberts is looking for someone to paint his house.

_______ **6.** Lisa can't decide if she should wear her raincoat.

_______ **7.** Kim wants to call a new friend, but she needs the phone number.

_______ **8.** Mrs. Hughes remembers to slow down her car near a school.

Name ______________________________ Date ______________

A Bar Graph

A **bar graph** uses bars to show information. The bars can be vertical or horizontal. The heights or lengths of the bars make it easier to compare data.

United States Ports That Export Goods

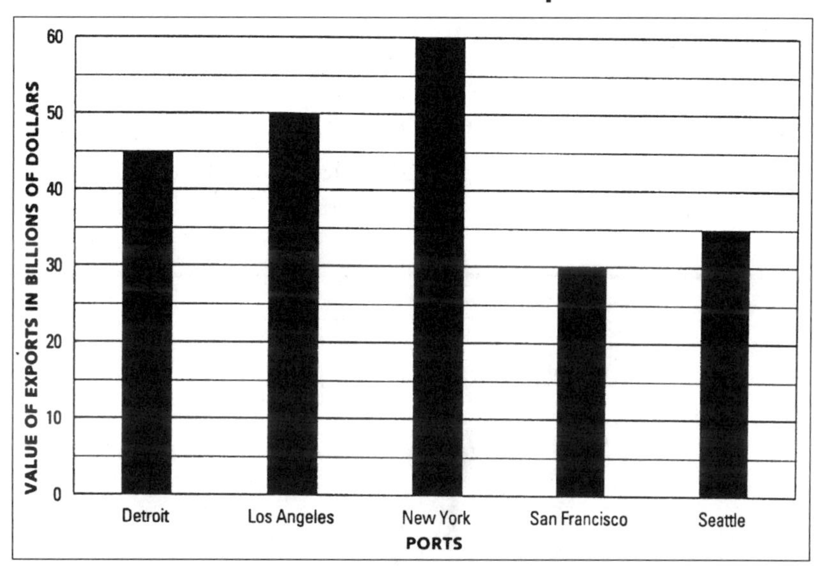

Directions **Look at the bar graph showing the United States ports that export goods. (An exported good is something that is made in the United States and moved to another country.) Answer the questions.**

1. Which U.S. port exports the most products?

____

2. What is the value of exports from San Francisco?

____

3. Does San Francisco or Detroit export the most products?

____

4. In dollars, how much more does New York export than Los Angeles?

____

Name ______________________ Date ____________

An Immigrant

An **immigrant** is a person who leaves one country to live in another country. People leave a country for many reasons. Long ago, the Pilgrims left their country and sailed across the ocean to America. They wanted freedom to follow their own religion.

When immigrants find a home, they often will eat the foods and use the speech of the country they left. They will even celebrate the holidays of that country. It is their **culture**, or ways of doing things. But they will also learn about the new country, too. They will learn the new language, try the new food, and enjoy the new holidays.

Directions **Write a paragraph explaining why people would move to a new country.**

____

____

____

____

____

____

Name ______________________ Date ______________

Regions: Unit 5 Assessment

Directions **Darken the circle by the correct answer to each question.**

1. Which of these could be a region based on a physical feature?

Ⓐ Jasper County Ⓒ the Rocky Mountains
Ⓑ Vermont Ⓓ Adams School District

2. Which of these is NOT part of a person's culture?

Ⓐ height Ⓒ foods
Ⓑ language Ⓓ holidays

3. How did the Northeast Region of the United States get its name?

Ⓐ The region is north of Canada.
Ⓑ The people living there are from the east.
Ⓒ It is named for a lake with the same name.
Ⓓ The states in it are in the Northeast part of the country.

Directions **Use the map to answer questions 4–6.**

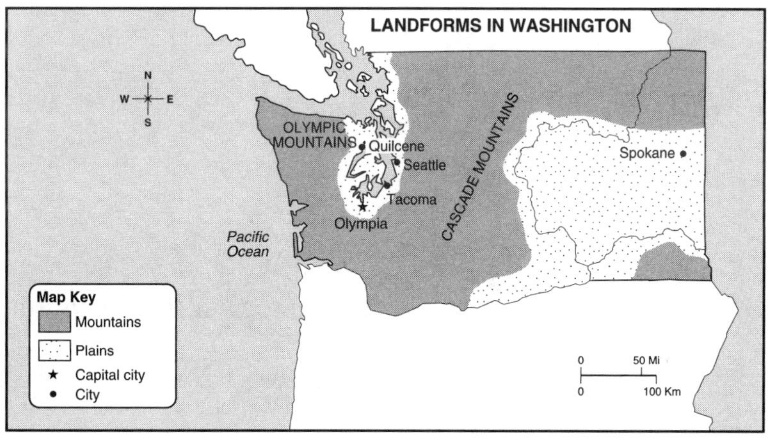

4. What kind of landform covers most of Washington?

Ⓐ mountains Ⓒ rivers
Ⓑ plains Ⓓ hills

5. In what landform region is the capital of Washington?

Ⓐ mountains Ⓒ rivers
Ⓑ plains Ⓓ hills

6. In what part of the state are most of the farms probably found?

Ⓐ north Ⓒ south
Ⓑ east Ⓓ west

Name ______________________ Date ______________

The Geography Theme of Regions

Regions are made up of areas that have something in common. Areas in a region can share physical or human features. Regions can be described as areas with the same climate, plant life, or government. Regions can be cultural areas where people share a language, food, or history. Regions can be as large as several states or as small as your community.

Directions **Look at the pictures. Answer the questions.**

1. What features tell about the region in this picture?

____

____

____

2. What features tell about the region in this picture?

____

____

____

Name ______________________ Date ____________

Comparing Parks

Regions are made up of areas that have something in common. Areas in a region can share physical or human features. A city is a kind of region. All cities have office buildings, apartments buildings, shopping areas, streets, and lots of cars and trucks.

Directions **Look at the picture of the park. Answer the questions.**

1. What features does the park in your community have?

____

____

____

____

____

2. How is a park a region?

____

____

____

____

____

Name ______________________________ Date ______________

Landform Regions

Regions are made up of areas that have something in common. Some maps show physical features like landforms, kinds of plants, or climate. It is easy to see how the different areas are formed into regions.

Directions **Use the map of North Carolina and Map E to answer the questions.**

1. What is the farm product raised closest to Winston-Salem?

____

2. Which three products are grown or raised near the town of Rose Hill?

____

3. Why are more crops grown in the eastern part of the state?

____

4. North Carolina and South Carolina grow the same kinds of crops and raise the same kinds of animals. Explain why both states are alike in this way.

____

____

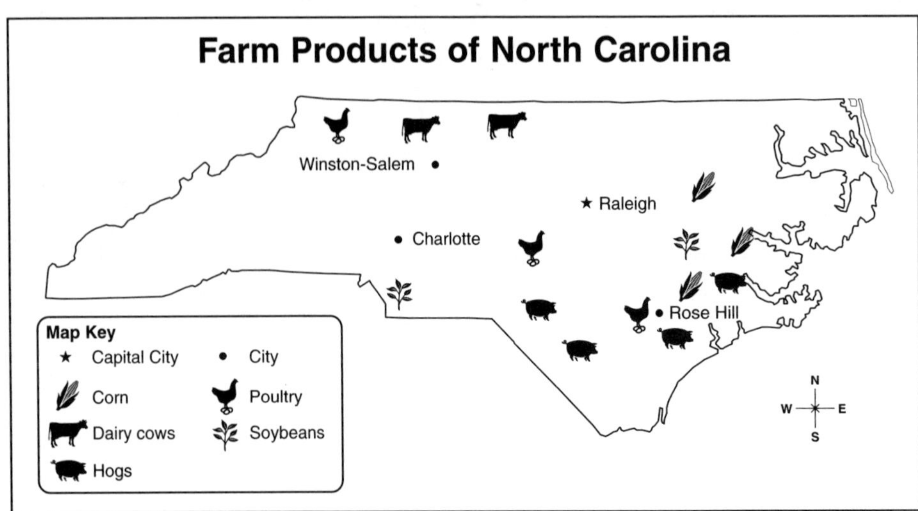

Name ______________________ Date ____________

Landforms and Physical Regions in the U.S.

Regions are made up of areas that have something in common. Places in a region can share physical or human features. A place can be described as being a part of several different regions. The United States is an example of how a place can be a part of different regions.

Directions **Look at Maps E and G. Answer the questions.**

1. What kind of regions are shown on Map E?

____

2. What kind of regions are shown on Map G?

____

3. How do you think the Pacific Region got its name?

____

4. What region got its name because of some mountains?

____

5. What four states are in the Southwest Region?

____

____

6. What kind of landform do all the states in the North Central Region have?

____

Name ______________________________ Date ______________

Cultural Regions

Culture is the food, language, holidays, and beliefs of a group of people. People who have the same culture often live near each other. They form a cultural region. In one community, people in different cultural regions might celebrate a holiday in different ways.

Directions **Read about some New Year's customs below. Then, draw a line from the group to the custom that matches that group.**

The first day of the calendar year is a holiday almost everywhere. In the United States, New Year's Day is January 1. Some people celebrate on a different day named by their religion. The Jewish New Year, called Rosh Hashanah (RAWSH huh SHAH nuh), is in September or early October. Hindus and Muslims celebrate New Year's Day on different days from year to year. The Chinese New Year begins between January 21 and February 19.

To celebrate the new year, people follow customs. For example, many people in the United States make resolutions. Resolutions are promises to act a certain way. In Belgium, children write messages on decorated paper for their families. In China, some people dress up like dragons at the end of a four day celebration. How do you and your family celebrate New Year's Day?

Group	**Custom**
1. Jewish people	Dress up like dragons
2. People in the United States	Celebrate *Rosh Hashanah*
3. Children in Belgium	Make resolutions
4. Chinese people	Write messages to family members

Name ______________________ Date ____________

Time Zones

When the sun shines on Earth, it is daytime on that side of Earth. But on the other side, it is nighttime. Because the Earth is so big, it is divided with imaginary lines to show different regions of time, or **time zones**. The time zones are numbered 1 to 24. All people in the same time zone set their clocks to the same time. As the number of the time zone rises by 1, the time becomes an hour later.

Directions **Look at the time zone map. Complete the table.**

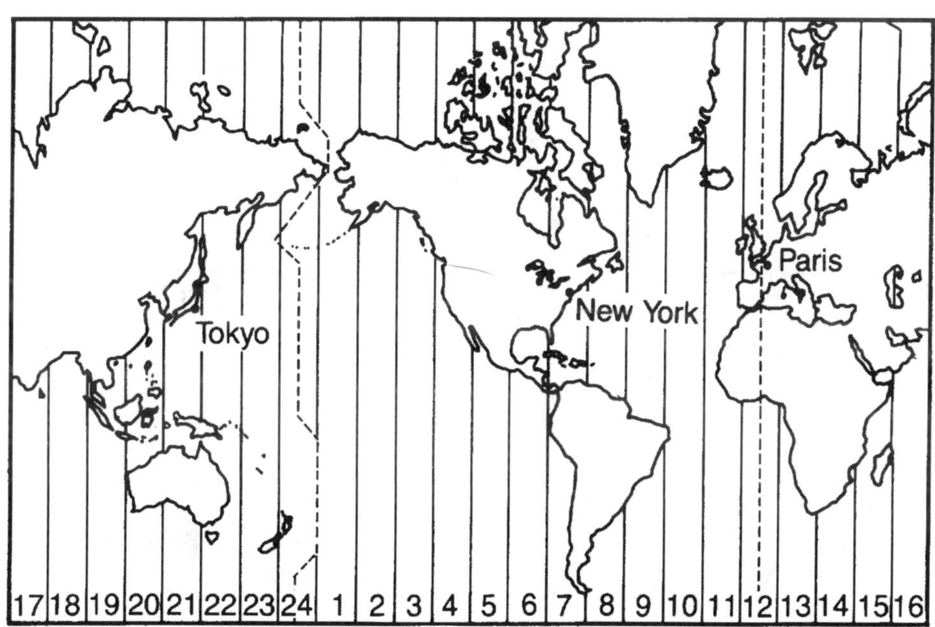

New York (zone 7)	7:00 A.M.	11:00 A.M.		5:00 P.M.	
Paris (zone 12)	12:00 noon		6:00 P.M.		
Tokyo (zone 21)	9:00 P.M.				2:00 P.M.

Name ____________________ Date ____________

Map A

Rob's Street

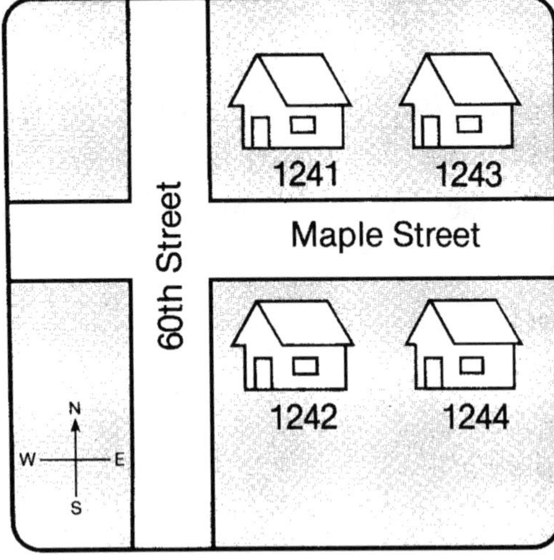

Map B

Maple Street Neighborhood

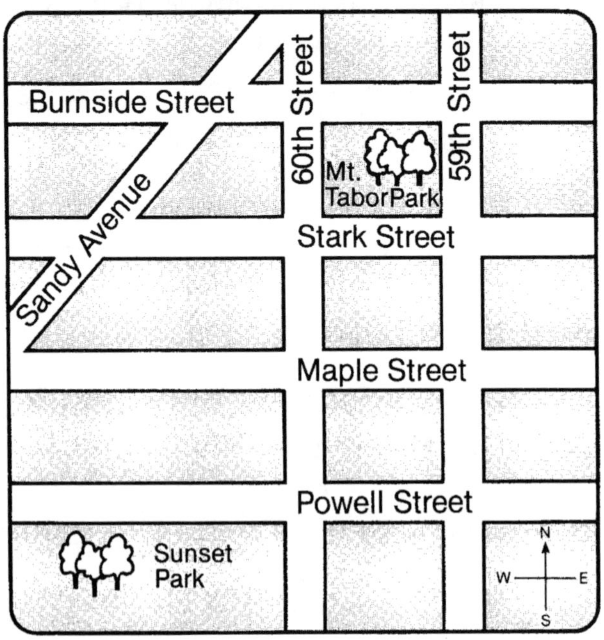

Map C

Bend River

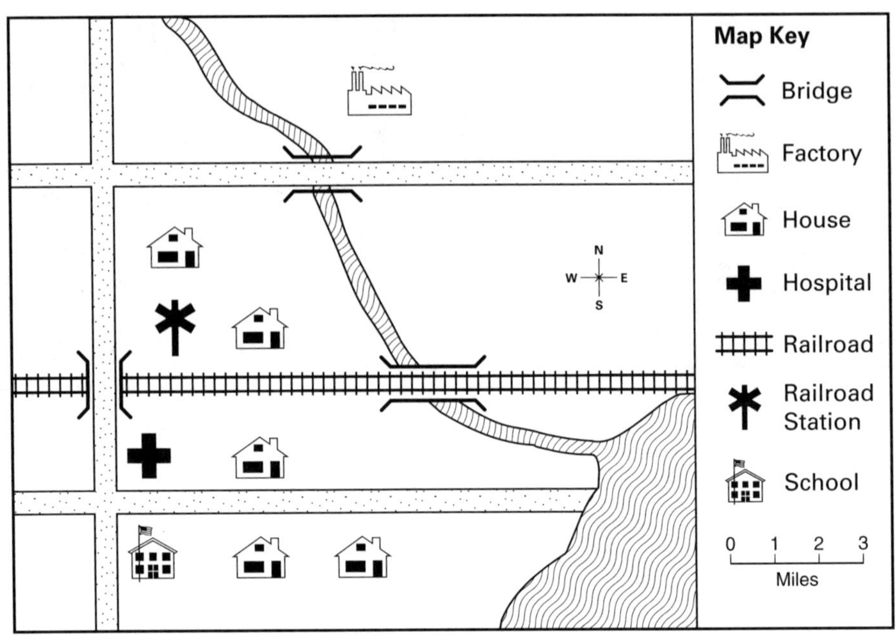

Name ______________________ Date ____________

Map D

Pennsylvania Relief Map

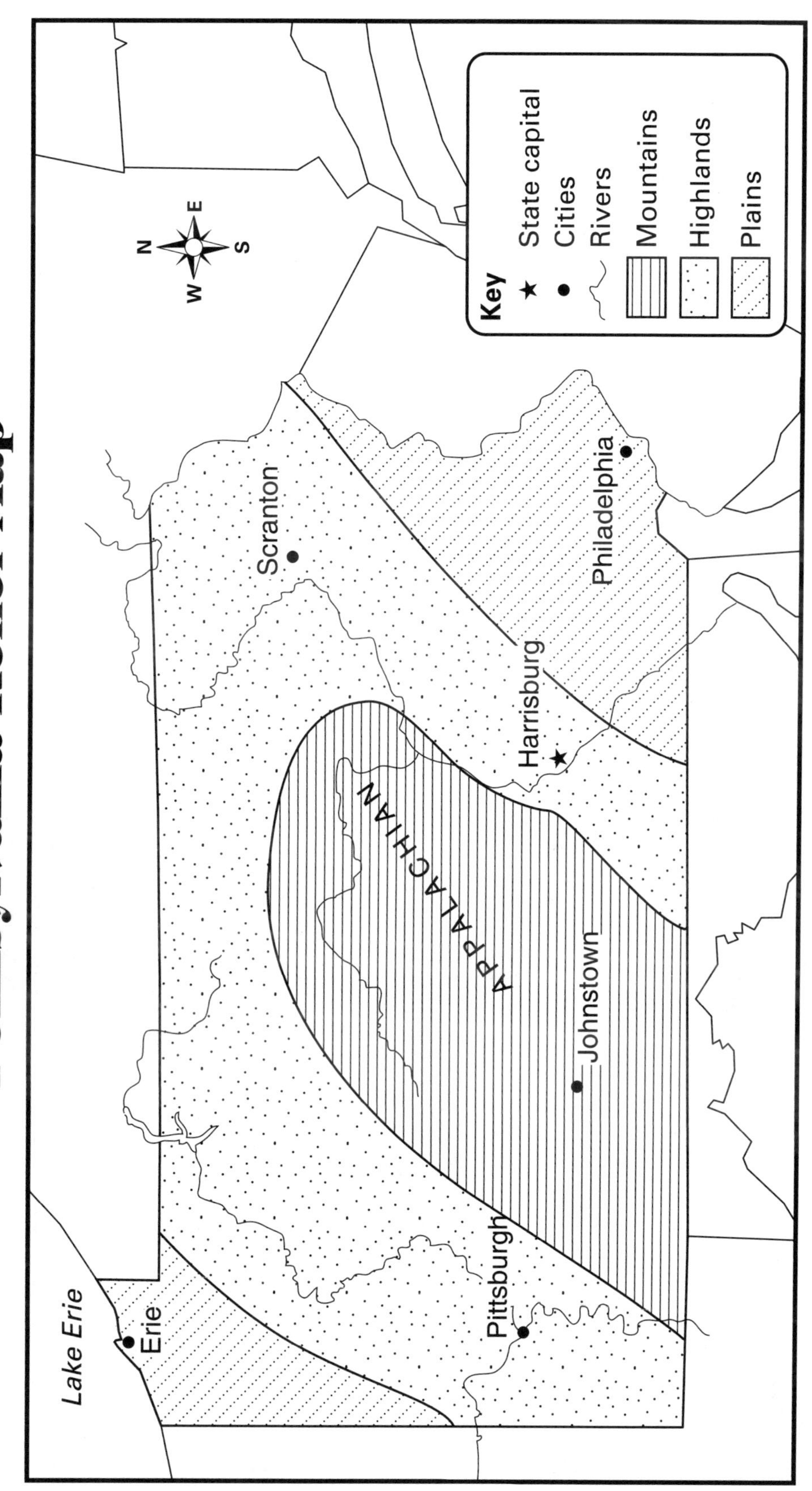

Name ______________________________ Date ________________

Map E

Landforms in the United States

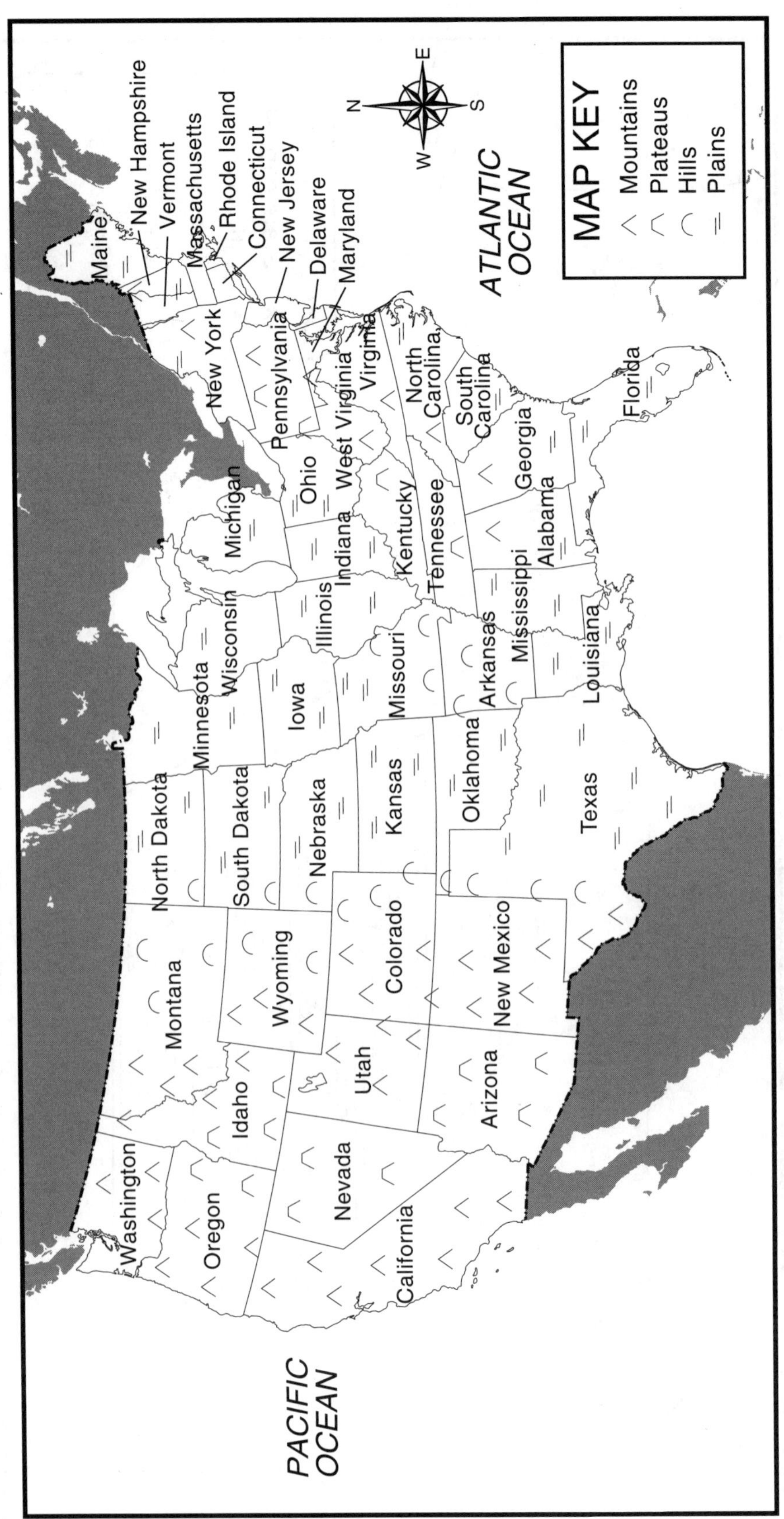

Name ______________________ Date ______________

Map F
United States
CANADA
WASHINGTON
Olympia
OREGON
Salem
IDAHO
Boise
MONTANA
Helena
WYOMING
Cheyenne
NORTH DAKOTA
Bismarck
SOUTH DAKOTA
Pierre
NEBRASKA
Lincoln
MINNESOTA
St. Paul
WISCONSIN
Madison
MICHIGAN
MICHIGAN
Lansing
CANADA
IOWA
Des Moines
ILLINOIS
Springfield
INDIANA
Indianapolis
OHIO
Columbus
CALIFORNIA
Sacramento
NEVADA
Carson City
UTAH
Salt Lake City
COLORADO
Denver
KANSAS
Topeka
MISSOURI
Jefferson City
KENTUCKY
Frankfort
WEST VIRGINIA
Charleston
VIRGINIA
Richmond
ARIZONA
Phoenix
NEW MEXICO
Santa Fe
OKLAHOMA
Oklahoma City
ARKANSAS
Little Rock
TENNESSEE
Nashville
NORTH CAROLINA
Raleigh
SOUTH CAROLINA
Columbia
GEORGIA
Atlanta
ALABAMA
Montgomery
MISSISSIPPI
Jackson
LOUISIANA
Baton Rouge
TEXAS
Austin
FLORIDA
Tallahassee
PENNSYLVANIA
Harrisburg
NEW YORK
Albany
VERMONT
Montpelier
MAINE
Augusta
NEW HAMPSHIRE
Concord
MASSACHUSETTS
Boston
RHODE ISLAND
Providence
CONNECTICUT
Hartford
NEW JERSEY
Trenton
DELAWARE
Dover
MARYLAND
Annapolis
Washington D.C.
MEXICO
Golfo de California
Pacific Ocean
Gulf of Mexico
Atlantic Ocean
N
W
E
S
Legend
★ State Capital
Juneau
ALASKA
Honolulu
HAWAII

Name ______________________________ Date ______________

Map G

Regions of the United States

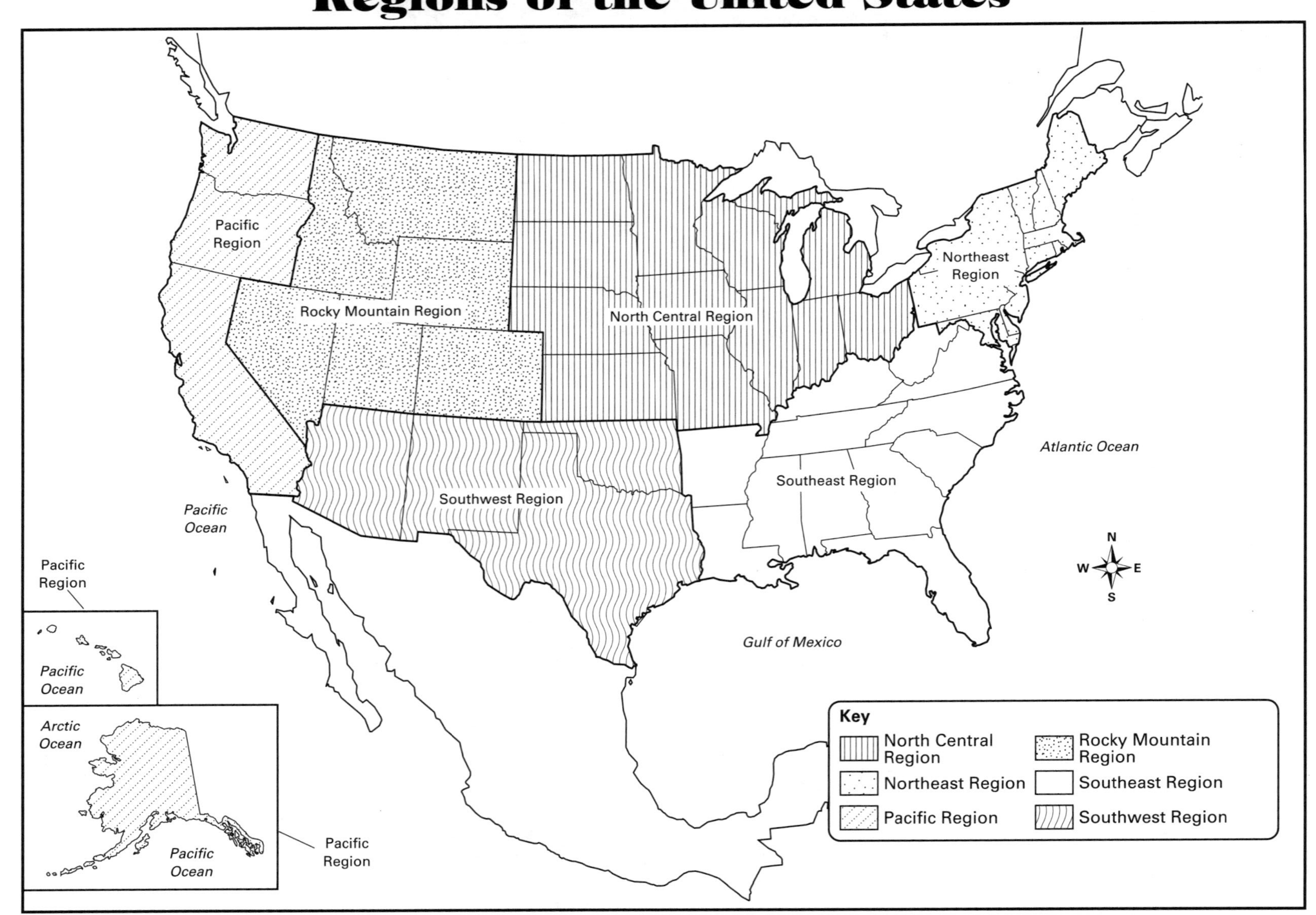

Geography, Grade 3

Answer Key

Page 3
1. B
2. C
3. A
4. A
5. C
6. A

Page 4

7. Possible answer: A warm climate all year long means that people can do more activities outside in the community, like go to parks.
8. Students name two of the following: college, schools, and libraries.
9. Accept reasonable answers.

Page 5
1. D
2. A
3. C
4. C
5. D
6. A

Page 6

7. ocean
8. Answer should be a job related to an ocean activity, like fishing or shipping.
9. A farmer would want to live on the plains because the soil is better for growing crops.
10. Areas that have mountains have tall areas of land, valleys, cooler climates, and fewer kinds of plants.

Page 7
1. B
2. C
3. A
4. D
5. D
6. C

Page 8

Possible answer: Elida lives at 123 South Street. Her house is back from the road in some trees.

Page 9
1. Students color the house with 1243 on Map A red.
2. Students color Maple Street on Map B orange.
3. Students color Mt. Tabor Park on Map B green.
4. Students color Oregon on Map F yellow.
5. Students color North Carolina on Map F blue.

Page 10
1. b
2. a
3. d
4. c
5. e

6.-10. Check students' maps.

Page 11
1. Carter Lake
2. B-2
3. Elmwood Park
4. C-2
5. Answers will vary.

Page 12
1. Philadelphia
2. Memphis and New Orleans
3. Pittsburgh
4. Students draw a mountain beside Denver.

Page 13
1. North America, South America, Antarctica, Europe, Asia, Australia, Africa
2. Atlantic, Pacific, Indian, Arctic
3. Rio de Janeiro, Sydney

Page 14

1. C	**2.** B
3. C	**4.** D
5. C	**6.** D

Page 15
1. Possible answers: barn, fence, truck, and office buildings.
2. Possible answers: mountains, stream, pond, tree, and plants.

Page 16

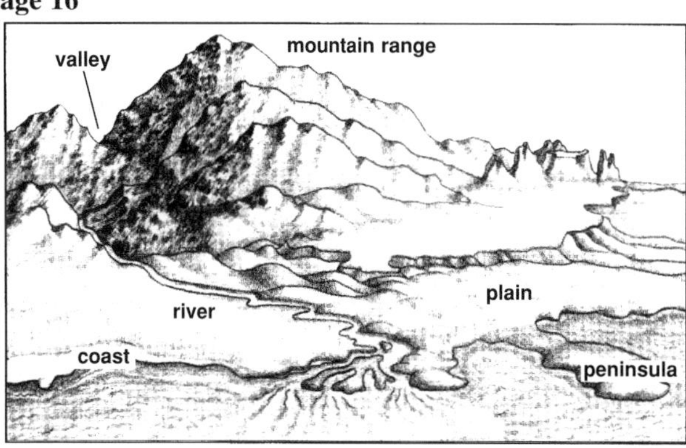

Page 17
1. Students mark the lake and all the trees with a P.
2. Students mark all the buildings and roads with an H.
3. Possible answer: There are roads, a fire station, a library, houses, and trees.
4. Accept reasonable answers.

Page 18
1. fishing; The Pacific Ocean is on the western side of California.
2. in the south
3. Students should name two of the following; gold, copper, zinc, forests, and cattle.
4. in the north; Factories often are built near the resource because the resource can be moved quickly and more cheaply.

Page 19
1. Accept reasonable answers, such as people do lots of outdoor activities and wear light-weight clothes.
2. Accept reasonable answers, such as people do lots of indoor activities or outside winter sports and wear warm clothes.

Page 20
1. River Thames and bushes and trees in Hyde Park
2. Students name two of these: buildings, streets, and bridges. Or students name two of the buildings or bridges shown on the map.
3. Accept reasonable answers, such as streets, buildings, and bridges.
4. Accept reasonable answers, such as buildings and sights special to London.

Page 21
1. C
2. A
3. C
4. A
5. B
6. D

Page 22
1. Answers will vary. Possible answers might include differences in weather, temperature, and terrain that would result in wearing different clothes, growing different crops, having different jobs.
2. People have cleared the

Answer Key

land to build office buildings and a house.

Page 23
1. 1993
2. The factory polluted the lake, and fish were killed.
3. Check that students write the events in the correct place on the time line.
4. Accept reasonable answers.

Page 24
1. coal, copper, natural gas, oil, and forests
2. forests and coal
3. forests
4. Craig, Gorman, Menlo, and Toler
5. Reed and Penn
6. Kelly
7. Accept reasonable answers that include why water is important, such as for drinking, transportation, and fun.

Page 25

Accept reasonable answers. Possible responses follow. Good Changes: New businesses are developed; roads are built; transportation improves; people have steady jobs. Harmful changes: The island becomes crowded; wildlife may lose their natural habitat; pollution increases; the island culture may be lost.

Page 26
1. Nevada
2. Dams hold water to keep areas from flooding downstream and to make lakes. Dams also generate electricity.
3. Custer Battlefield

Page 26 cont.

4. Answers will vary, but should say that it was the site of an important battle—specifically, the battle in which General George Custer's army troops were defeated by many Native American groups.
5. Answers will vary, but should say it needed to be made into a park to keep it safe for many people to see and enjoy.

Page 27

Possible answers: The animals will die without their homes and food. The cultures of the different tribes will be lost. The medicines made from plants cannot be made anymore. There will be a lack of rubber for different products and a lack of other materials for goods in the home.

Page 28
1. D
2. B
3. A
4. D
5. A
6. C

Page 29
1. Goods are being moved by a ship.
2. Ideas are being moved by a person reading from a book.

Page 30
1. streets and train tracks
2. Students should name three of the following: school, library, museum, and post office.
3. Answers will vary.

Page 31
1. Interstate 40
2. about 120 miles
3. south, then west

Page 32

Answers may vary.
1. a, b
2. b
3. d
4. c
5. a, d
6. a, b
7. d
8. c

Page 33
1. New York
2. $30 billion
3. Detroit
4. $10 billion more

Page 34

Possible answers: jobs, education, adventure, better way of life, freedom of religion, escape from war, or more food.

Page 35
1. C
2. A
3. D
4. A
5. B
6. B

Page 36
1. Answers will vary but should include references to a desert environment.
2. Answers will vary but should include references to a cultural area.

Page 37
1. Answers will vary but should include references to playground equipment, plants, and activities.
2. Answers will vary but should include references to playground equipment, plants, and activities.